Why Would Anyone Cut a Tree Down?

Why Would Anyone Cut a Tree Down?

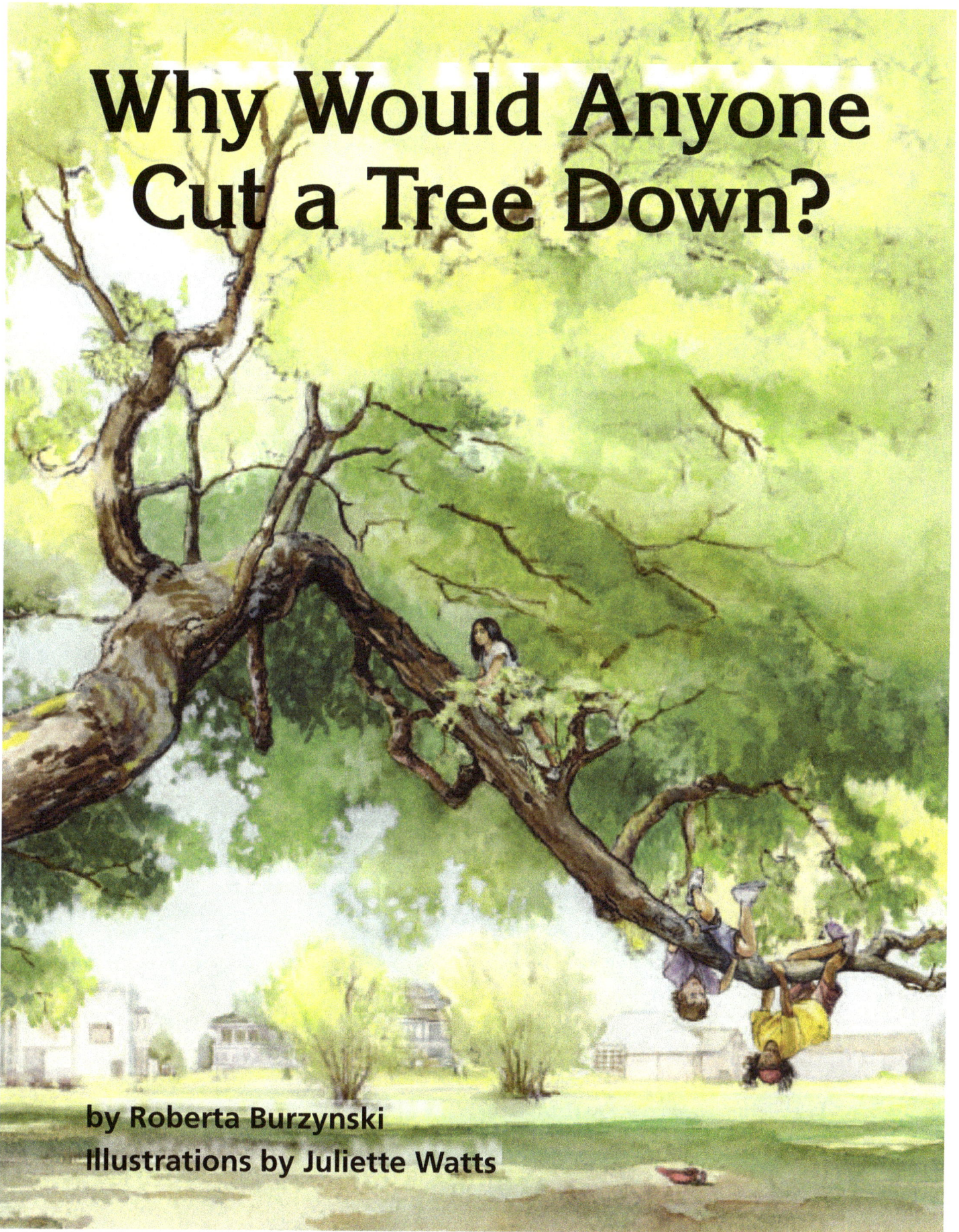

by Roberta Burzynski

Illustrations by Juliette Watts

Preface

While giving presentations about trees, I have been impressed by students' knowledge of the benefits and products that trees provide. On the other hand, I have been surprised when students blurt out that you can never cut a tree down. To bridge this gap in knowledge and understanding, I wrote this book.

In taking care of forests, towns, or backyards, it is sometimes necessary and even helpful to cut trees down. This book is meant to introduce everyone from elementary students to adults to different reasons that foresters and arborists might cut trees down.

Sometimes a tree needs to come down because people did not care for it properly. So, the back of this book includes some basics of tree care and places to get more information.

I hope this book helps you see that, in caring for forests and trees and meeting people's needs, cutting a tree down is not always a bad thing.

—Roberta Burzynski

Acknowledgments

A world of thanks goes to the many people, too numerous to name, who helped with this book in a variety of ways—developing the concepts, reviewing drafts, ensuring accuracy of the text and illustrations, planning the marketing, arranging for printing, and creating the related materials and Web site. Your time and talents are very much appreciated. Last but not least, we both thank Roberta's aunt for her constant support and gentle encouragement that kept us going.

—R.B., J.W.

USDA Forest Service

Northeastern Area State and Private Forestry
11 Campus Boulevard, Suite 200
Newtown Square, PA 19073

www.na.fs.fed.us

NA–IN–01–12
February 2013

USDA is an equal opportunity
employer and provider.

Why would anyone cut a tree down?
Trees do terrific things
for the land and air,
water and wildlife,
and for you.

Trees give you shade and oxygen.
They soften loud noises.

Trees slow the
wind and keep
the soil from
sliding away.

They clean the
air and water.

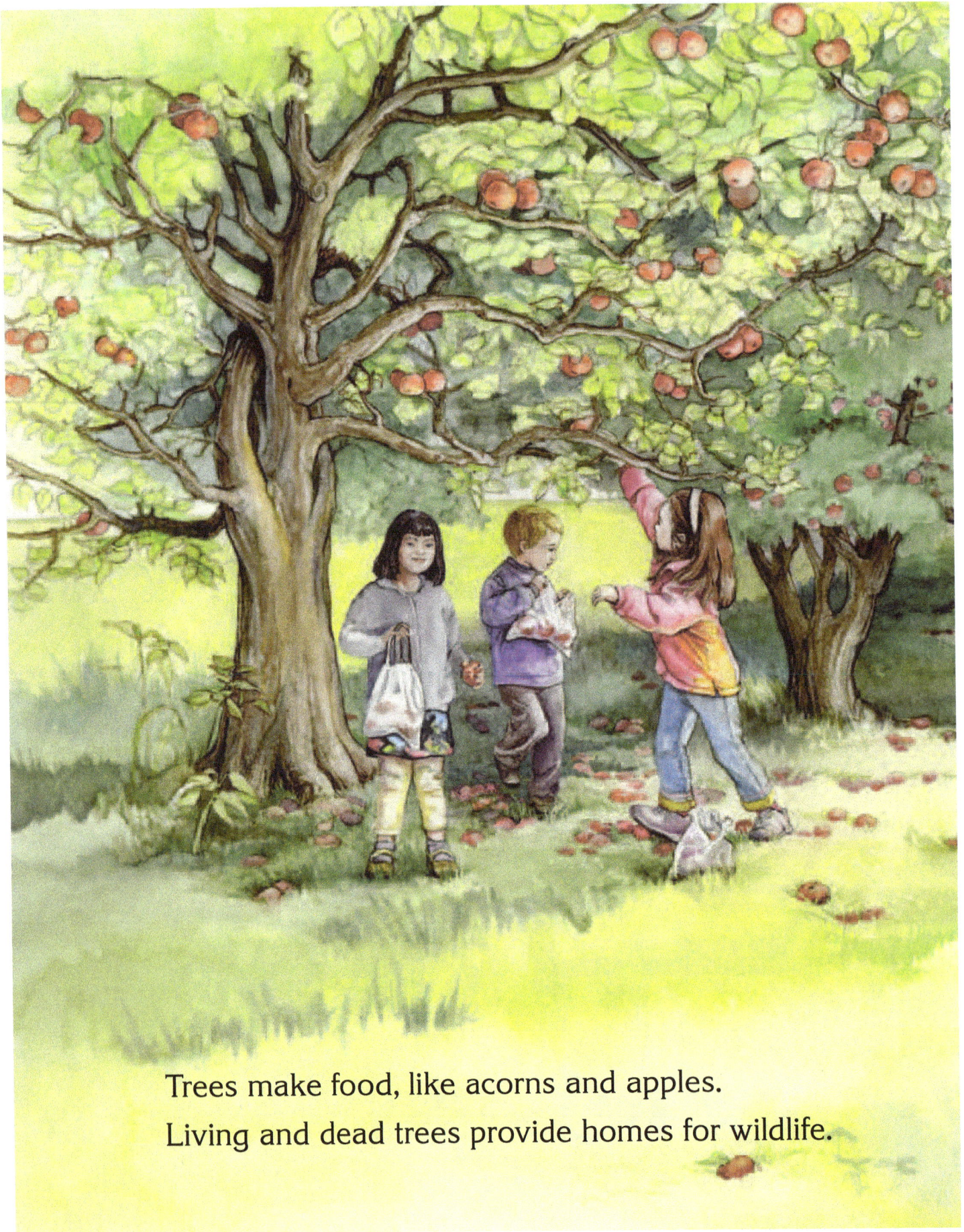

Trees make food, like acorns and apples.

Living and dead trees provide homes for wildlife.

Seeing trees makes you feel better.

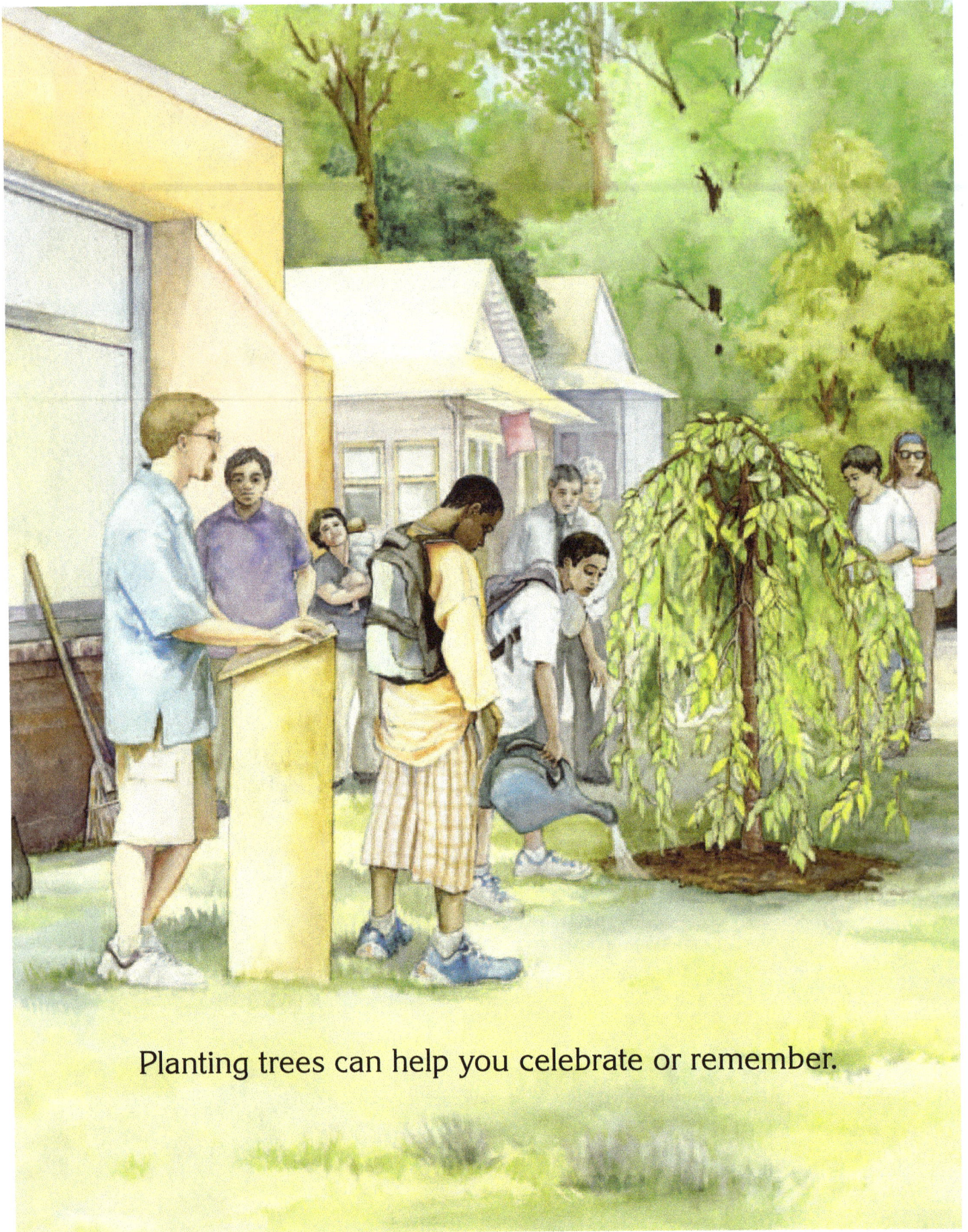

Planting trees can help you celebrate or remember.

So, why would anyone cut a tree down?

Trees grow. Trees can get sick
and develop problems. Trees get
old and die, like all living things.
And trees are made of wood.

So, to take care of forests
and towns, to keep people safe,
and to get wood, people cut
some trees down. We may not
want to, but sometimes we need to.

Some trees have defects or become damaged.
Defective and damaged trees can be dangerous.
Taking down dangerous trees keeps you safe.

Some trees get sick. Sick trees may make other trees sick too.

Removing sick trees may help other trees stay healthy.

A forest of trees can catch and spread fire.
Trees that are too close to a house in the woods
can create a fire hazard.

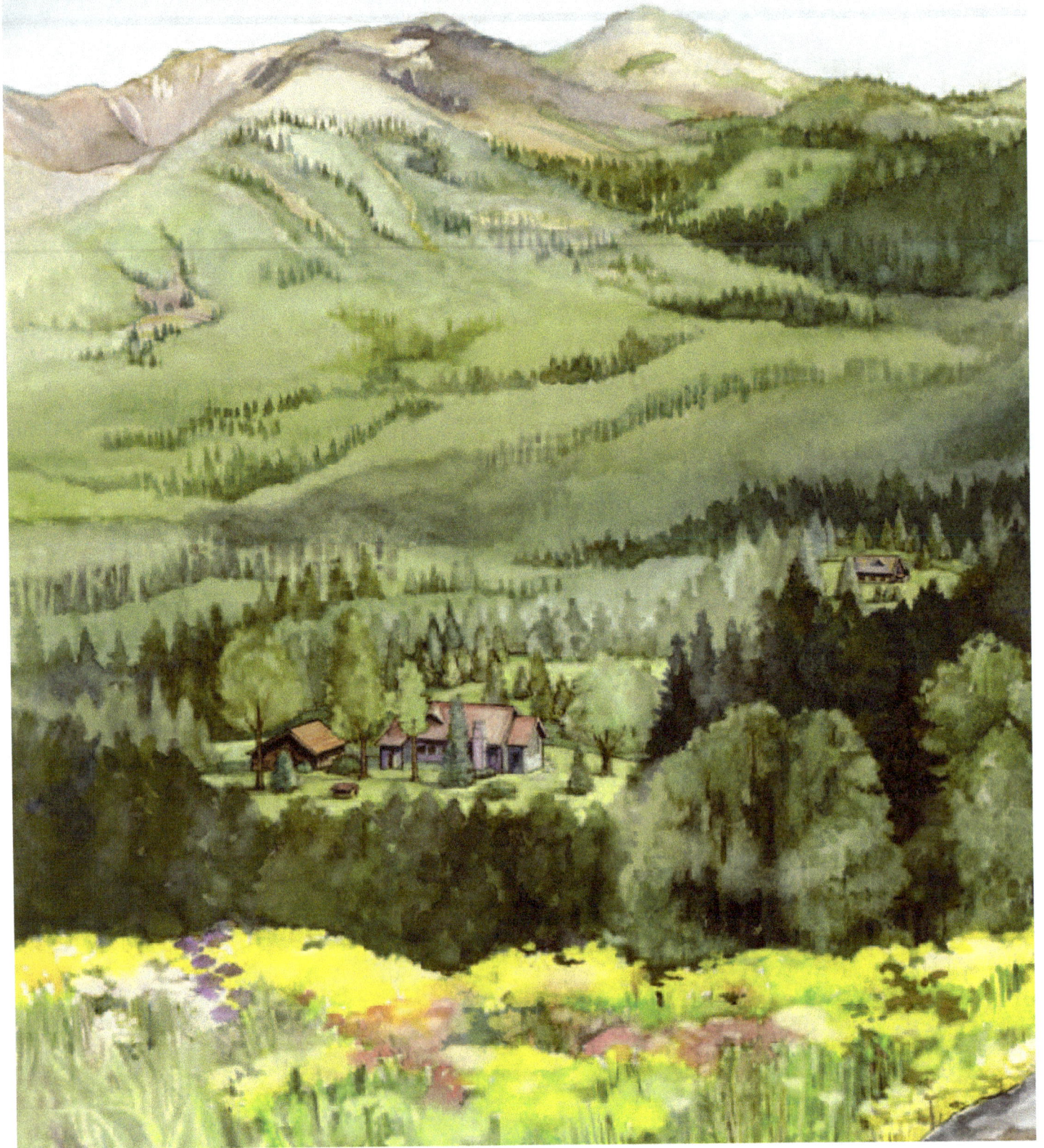

Removing some trees helps to keep a woodland home safe from wildfire.

Some trees grow too close to one another. They may be so close that none of them gets enough sun, water, or nutrients. Plants underneath the trees may not grow well.

Removing some trees champions other trees and gives them more room to grow and thrive. Forests are healthier and produce more food for wildlife.

Trees are made of wood, and wood is made into things you use.

Harvesting some trees allows you to have things made from wood.

People need to cut some trees down,
but that is not the end of the story.

Things made of wood can last a long time,
and you can use them over and over again.

Insects and other organisms turn toppled trees
into new soil that helps other trees grow.

Some tree stumps send up new shoots and grow again.

Some tree seeds sprout into new trees.

You can plant a new tree and
take good care of it.

Trees are renewable!

A Note to Parents and Educators About Caring for Trees

Sometimes a tree needs to come down because people did not care for it properly. So here are basics of tree care, and resources for more information.

Taking down dangerous trees

There are several signs that a tree is getting weak and may fall—dead wood, cracks, decay, cankers, root problems, poor form, and weak branch unions with the trunk. These problems are shown in How to Recognize Hazardous Defects in Trees, by the Minnesota Department of Natural Resources and USDA Forest Service (http://na.fs.fed.us/pubs/detail.cfm?id=927). Even trees with these defects are not considered dangerous, or hazardous, unless the falling tree would hit a building, vehicle, or place where people often gather or walk. Pruning the tree may correct the problem, or the tree may need to be cut down. Many hazardous defects can be avoided by correctly planting and pruning young trees. For proper pruning techniques, see How to Prune Trees, by Bedker, O'Brien, and Mielke (www.na.fs.fed.us/pubs/detail.cfm?id=2602); and the animated Tree Pruning Guide on the Arbor Day Foundation Web site: www.arborday.org/trees/pruning/.

Removing sick trees

Insects and diseases that were accidentally introduced from other countries can cause serious tree health problems. Because these pests have no natural enemies in this country, they can multiply rapidly and damage trees. Removing and destroying infested trees may reduce the numbers of pests and keep them from attacking other trees. Oak wilt disease, Dutch elm disease, and Asian longhorned beetle are some of the pests that make it necessary to remove infested trees. Not all infested trees need to be cut down. Insects and diseases that are a natural part of a forest usually do not cause problems as serious as those caused by introduced pests; however, even native pests, like bark beetles, can suddenly increase in number and kill trees. For help identifying tree pests, contact your State Forestry office or your county Cooperative Extension Service.

Removing flammable trees

Trees, shrubs, and ground cover plants provide fuel that can feed a wildfire. If a home is surrounded by flammable vegetation, removing the vegetation creates a safety zone that helps protect the home from wildfire. A building on a steep hill needs a wider safety zone than one on a gentle hill or flat land. Many plants naturally resist fire and will keep it from spreading. For more information on using fire-resistant plants in Firewise landscaping, visit the National Fire Protection Association Web site: www.firewise.org.

Cutting down crowded trees

In the woods, if the tops of trees touch, then they are competing with each other for sunlight, water, and nutrients. And if the trees are over 25 feet tall, cutting some of them will help the remaining trees grow. A forester should decide which trees to cut. The cut trees can often be used for wood products. If the wood cannot be used or sold, then there are a couple of other options. The trees can be killed by girdling (cutting into the bark in a continuous band around the tree trunk) and left standing for wildlife homes, or they can be cut down and left on the ground to decompose and recycle their nutrients. In cities and towns, trees may outgrow their planting space and interfere with buildings, utilities, driveways, or walkways. To learn the size of trees at maturity before you plant them, use the Tree Guide (www.arborday.org/trees/treeguide/index.cfm) or the Right Tree, Right Place Guide (www.arborday.org/trees/righttreeandplace/), both on the Arbor Day Foundation's Web site. Be sure to pick the right spot and the right tree, then plant it right, as shown in the planting diagram on page 40.

Using wood from cut trees

Trees that need to be cut because they are crowded, sick, or pose a hazard can often be used for wood products. The most common products are veneer, boards, fence posts, and firewood. Black walnut, red and white oak, sugar maple, and black cherry trees with a straight trunk 20 inches or more across can provide thin sheets of veneer for furniture. Trees with a straight trunk 12 inches or more across can be cut into boards. Trees that are naturally resistant to decay and insect attack—such as cedar, cypress, redwoods, and white oak—make good fence posts. Trees that are crooked, badly shaped, infested with pests, crowd more desirable trees, or have died and fallen to the ground are good for firewood. Local artists may be interested in unusually shaped logs or stumps that can be made into unique wood creations. Remember not to move firewood over distances, so you do not spread invasive insects that may be in the wood. To locate wood markets besides commercial sawmills, contact the State Forestry Office, manufacturers of portable sawmills, State-wide craft organizations, county Cooperative Extension Service, or Resource Conservation and Development (RC&D) council.

Much of this information was taken from these Tip Sheets: Help your preferred trees grow, Identify and manage hazardous defects in your trees, Keep your woods healthy, Protect your property from wildfire, Generate wood products. The National Arbor Day Foundation. Nebraska City, NE. www.arborday.org/backyardwoods/ *(link: Download tip sheets) (1 November 2010).*

Pick the right spot and the right tree, then plant it right.

(1) Choose a site and consider the sun exposure, moisture, soil, wind, and salt exposure.

(2) Choose a tree that is suited to the site and will have enough space for its branches and roots when fully grown.

(3) Dig a hole 2–3 times wider and no deeper than the roots. Expose the root collar— the place where the trunk meets the roots. Remove container or rope and wire basket, and cut off any roots that circle the root ball.

(4) Place the tree in the hole and push the top of the burlap down so it rests on the bottom of the hole. Put the soil back in, up to the level of the root collar.

(5) Add mulch 2–3 inches deep over the planting area making sure it does not touch the trunk.

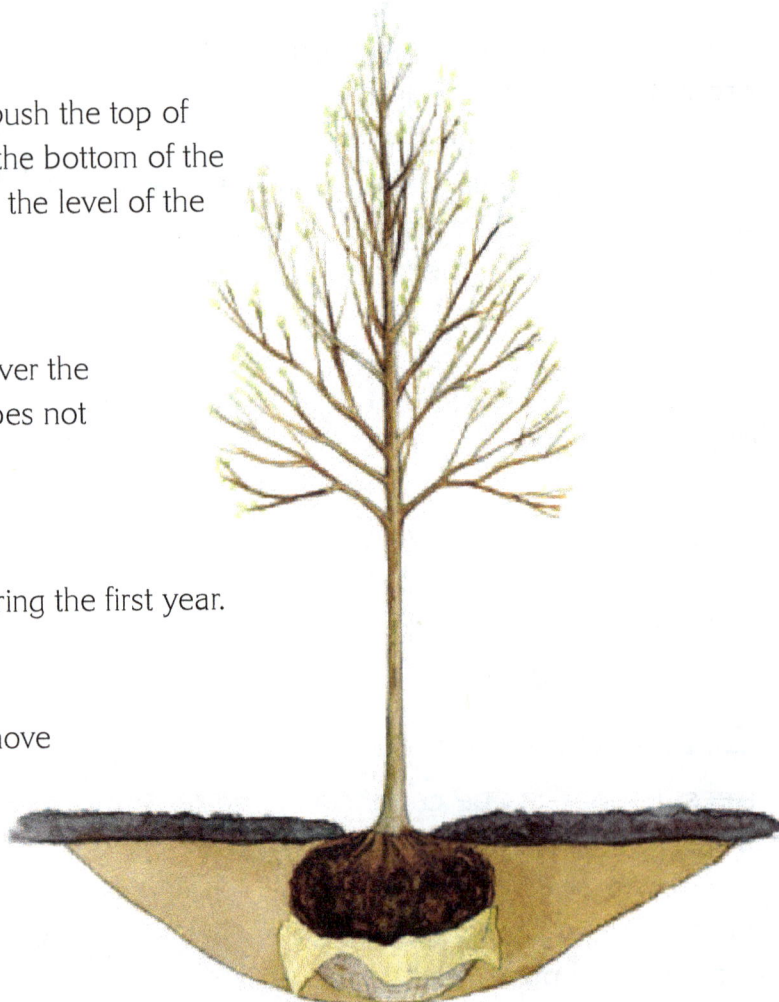

(6) Water deeply and regularly during the first year.

(7) Prune only if necessary to remove damaged branches.

Planting to Remember

Rod Whiteman and Dan Snider

Family, friends, and coworkers planted 215 trees in memory of Forest Service workers, Rod Whiteman and Dan Snider, who—along with pilot Patrick Jessup—lost their lives on the job, on June 21, 2010. This planting at Fort Necessity National Battlefield in Fayette County, PA, will serve as a memorial to these dedicated public servants, for decades to come.

Kathy Cellucci

The sixth grade class at Tatem Elementary School in Collingswood, NJ, lost their beloved teacher, Mrs. Kathy Cellucci, in November 2007. The following spring, students, family, and friends planted a tree on the school grounds to honor her. Pages 12 and 13 in this book capture that memorial gathering.